Top 10 Vaccine Objections

Top 10 Vaccine Objections

Objections

Doubts and Conversations

ALEX RAMIREZ

EDITED BY CHARLES NATON

ANTHEM PRESS

UNION BRIDGE BOOKS
An imprint of Wimbledon Publishing Company Limited (WPC)
UNION BRIDGE BOOKS
75–76 Blackfriars Road
London SE1 8HA
www.unionbridgebooks.com

A CIP record for this book is available from the British Library.

ISBN-13: 978-1-78527-539-5 (Pbk)
ISBN-10: 1-78527-539-9 (Pbk)

This title is also available as an e-book.

CONTENTS

ACKNOWLEDGEMENTS

Many thanks to Paco and Susana for the wonderful illustration work.

INTRODUCTION

Hi! My name's Alex and I'm a biochemist.

'I'm a biochemist' is a pretty short and succinct response to the inevitable occupation questions that come up within the first half minute of meeting somebody new.

It's not just me that it happens to, as most of us end up talking about our work in social situations because people are naturally curious. That's why the fireman is asked about running into burning houses and the dentist ends up fielding questions about that toothache that kind of comes and goes. We all wonder about the lives of others, especially when someone has an unusual kind of job.

It's easy for me to forget just how unusual my job actually is, and I suspect that's largely because I spend most of my time locked in the lab with other biochemists, along with some other assorted flavours of nerds. However, I'm reminded of the fact that I do have an unusual occupation when so many people ask me to tell them a little more about what I actually do with my day.

My pleasure, I'm glad you asked.

I'm a biochemist with a PhD in molecular immunology. When people ask me to translate that into English I explain that I do a lot of work with vaccines, both in academia and in the commercial sector. Now if it's a good party and I'm feeling three glasses of wine grandiose (and my conversation partner hasn't already fallen asleep), then I might expand my explanation still further to cover my specialist areas of antigen discovery, adjuvant research and process development. When I'm inevitably asked to translate that into English, I just explain that I spend my time trying to figure out exactly which components are essential for a vaccine to make it work. I also do a lot of research into figuring out whether there's anything we can add to an existing vaccine to improve it, as well as helping to develop efficient, safe and cost-effective production techniques for the pharmaceutical sector.

The end result of all that fancy science-speak is that, in terms of vaccines at least, not only am I the guy who sees how the sausage gets made, I'm actually the guy who comes up with the recipe in the first place.

Yeah, that's me, the vaccine sausage guy; although that's not how it's written on my business card.

Like I said, I have a pretty unusual job and people are often very interested when they realise they're talking to a genuine, fully paid-up member of the bona fide expert club – at least when it comes to vaccines. Sausages, less so.

When I'm not drinking wine at parties or talking to other scientists about vaccines, I spend a lot of my time talking to doctors, students and members of the public about the same subject, so I guess that really makes me a kind of go-to guy when it comes to questions about vaccination.

That's how you came to be reading this book.

A quick Internet search will immediately tell you that there are plenty of other books about vaccines available right now. There are lots of incredibly detailed academic works, professional journals and epidemiology studies – all outlining the case for using vaccines and providing the data to back it up. So much data . . .

However, there are precious few books written by insiders like me that aren't three feet thick and crammed with unintelligible graphs, tables and footnotes that in many cases are even longer than the text itself.

There are no graphs in this book, just as there is no unnecessarily complex language and no soul-destroying swamps of appendices and references. Instead, this is just a plain English summary of the huge number of discussions I've enjoyed with folk from all walks of life about the subject of vaccination. You'll be happy to hear that I favour the amusing and the informative rather than any preachy, finger-wagging medical lectures. Many people are concerned about the perceived efficiency, safety or even the morality of vaccine use, and although it's not technically part of my job to answer those questions, I do have some special insight regarding these matters.

My conversations have happened in bars, restaurants, aeroplanes and even in elevators. To be honest, I really don't know how many hundreds there have been over the years, but they can be boiled down to 10 basic objections to vaccination on various grounds. So if you're worried, concerned or just want some reliable yet casual information about vaccines, this is probably the best place to find the answers you're seeking.

It's completely natural to have questions and doubts, especially with so much lurid and frankly wrong information hurtling around in cyberspace, but I'm here to just give you the facts in a straightforward manner that doesn't need a science or statistics degree to interpret. I have no reason or desire to make anyone feel guilty, or to shove my point of view down anyone's throat. My aim here is to provide a different, relaxed and yet well-informed perspective. If some readers experience a sudden 'I never thought about it like that' moment at the end of it all, then I'll know I've done my job well.

Anyone can enjoy this book, including other professionals who may have a solid understanding of vaccines and the data surrounding them, but who struggle to communicate that knowledge to the vast majority of people who don't have any training in epidemiology or immunology.

There aren't that many of us professional vaccine immunologists around out there, and we can't talk to everyone in person. This book is the next best thing to a face-to-face conversation with the guy who understands vaccines so well that he actually helps to cook up the ones we already have, as well as working to develop new ones for the future.

Let's start the conversation.

I DON'T BELIEVE IN VACCINES

I hear this sort of generic statement a lot, quite often as a response when I tell people what I do for a living. Assuming I'm not immediately banished to social perjury before the conversation's begun, I usually start by enquiring as to why someone doesn't believe in vaccines.

What most people really mean when they say that they don't believe in vaccines is that they don't believe vaccines are necessary, or effective, or beneficial to health.

If that's the case then we're usually off and running by that stage, as I'm fairly certain I'm going to hear some variation of the arguments we'll be exploring in the following chapters.

I'm not lucky enough to possess any kind of magic crystal ball for seeing the future, but having spent most of my adult life talking about vaccines in one way or another, I have a pretty good idea of what's coming next when the issue arises. Those issues could be safety concerns, the autism scare or some other subject area where I can offer my expertise and hopefully some reliable reassurance that ultimately there's really not that much to worry about.

However, I do occasionally come across an argument that's far beyond my sphere of influence. When that happens, I've learned it's time to beat a polite retreat because I already know there's no chance of enjoying a discussion based on mutually recognised facts and values. It's a bit like two people talking in different languages. For example, if it's a purely religious objection then I've

learned that it's a really bad idea to discuss the issue of vaccines in theological terms . . . unless you're a priest. In truth, I've got no idea what Jesus might've thought about vaccines, so I can't really talk about the issue in theological language. In that kind of situation I know it's time to admit defeat because those arguments are simply outside my area of competence.

Although it may be true that civility is society's engine oil, I'll admit that sometimes I can't resist a parting shot like, 'It's okay not to believe in vaccines. After all, viruses don't believe in you either.'

To be honest, I doubt if that kind of pithy comeback has ever persuaded any religious objector to look at the vaccine issue in a different way, but then it's really more for my own amusement rather than to convince anyone of anything. I just find it funny because not believing in vaccines as such is similar to not believing in viruses or bacteria; indeed, some vaccines are actually types of viruses or bacteria.

Although mercifully rare, I have met some pretty colourful characters during the course of my career, and I tend to disengage as quickly as possible if I'm unlucky enough to get caught in their orbit. You know the kind of people I mean: the ones who genuinely believe that aliens were responsible for the JFK assassination and think that the 9/11 attacks were perpetuated by reptoids. If someone's willing to dismiss any and all available evidence as just another fabrication by some all-powerful yet completely invisible and untraceable state apparatus, then they're far too 'smart' to believe a single word I might have to say.

In this admittedly unusual situation, I know I'm not dealing with an entirely rational kind of mindset, so I immediately understand that I'm spinning my wheels and wasting my breath.

That's when I make a strategic withdrawal to the bar or the cloakroom ... or anywhere really, just to put a safe distance between us and to ensure that my secret government mind-warping device won't jangle their tinfoil hat and give the game away.

While it's sadly true that some folk just aren't built for rational discussion, there's one subject that's pretty much guaranteed to make all of us into crazy people.

We're all hardwired to be highly irrational when it comes to our children, and when you think about it, that's probably necessary for survival. After all, a parent may be called upon to put themselves in danger to protect their offspring or, in the most extreme cases, sacrifice themselves to preserve their children.

I immediately know I'm walking blindfolded through a minefield whenever the subject of what's best for somebody else's child comes up. Now I don't know about you, but I'm not all that keen on tiptoeing my way through acres of explosives when I know I don't have to.

Of course, when I'm discussing childhood vaccinations, it's really the parent's objections I'm addressing and not the child's. Convincing a toddler to get a jab is as easy as offering the promise of a lollipop afterwards, as any doctor throughout the world will tell you.

Experience has taught me that first-time parents are the most anxious group when considering vaccinations for their children, despite the fact that they're quite likely to have undergone those very same immunisations when they were kids. This is actually rather helpful because it's far less emotive to talk about adults than infants in this kind of situation. If an adult's already been vaccinated, then it's much easier to highlight the enormous health benefits they're enjoying at every minute of every day, and just remind them that their kids can also enjoy those same benefits at a minimal risk.

This fundamental rule holds true for all of the discussions that are to follow. If someone's talking about kids, then I talk about adults. It's the quickest and easiest way to take most of the heat out of the encounter and focus on the real issue at hand. Generally speaking, if an adult can be persuaded that vaccines are a good thing for them, they'll automatically decide they're a good thing for their kids as well.

1

I NEVER GOT THEM AND I'M FINE

This is something I hear sometimes, and it often comes as part of a package which can include the relative in his 90s who smokes twenty a day and the health freak who got hit by a bus. For some reason, a bus seems to be the most dangerous vehicle out there. Maybe someone should look into that.

This kind of unvaccinated and generally lucky person seems to believe that because nothing bad has happened to him in the past, he doesn't need to take precautions against the future. In fact, he's so confident in his good fortune that he sometimes wonders whether vaccines serve any useful purpose at all, other than to offer some kind of empty reassurance.

Johnny Goodluck's entire ethos seems to revolve around the idea that precautions make very little difference if your number's up. There's almost an element of stoic fatalism in this kind of outlook, but it's usually pretty superficial when tested in the real world.

Years of making very little progress in situations like this has taught me that it's far better to discuss vaccination in terms of things that we can all relate to, rather than trying to explain with hard data, facts and figures.

I'm not really surprised that the average, reasonably healthy 40-something sees very little reason to go rushing off to the doctor because he missed a few injections when he was a toddler. Those missed appointments were many decades ago and, assuming he lives in the developed

world, it's doubtful he's had much if any contact with victims of poliomyelitis or whooping cough. He hasn't been vaccinated, and as yet he hasn't contracted any of the diseases that his non-existent injections are designed to protect him against. If I'm honest, there's a good chance he could live out his entire life without ever catching measles or mumps, much less polio. The real reason he doesn't encounter these diseases is because they've been virtually eradicated in the industrialised world. So why should he bother so late in the day?

Well, simply put, he should bother because vaccines are like seatbelts, and the fact that you've never had a car crash before doesn't necessarily mean you won't have one tomorrow. If you never wear a seatbelt, you might have concluded that it's a useless car safety feature . . . so long as you never have a traffic accident.

However, it's an inescapable truth that accidents can and do happen to the good, the bad and even to the lucky. You could be the most careful and considerate driver in the entire world, but that doesn't mean the other guy at the junction isn't drunk, half asleep or just not paying attention. You could be doing everything right all the time and still end up with your door dented in or, God forbid, something much more serious. The fact that it's not your fault won't make your car any less written off or you any less injured.

Now just imagine being that same careful driver who's been on the road for 20 years without ever wearing a seatbelt! I guess you don't need me to elaborate on what could happen if that other drunk or sleepy guy ploughs into you when you're not safely tethered inside your car. It could literally mean the difference between being pretty badly shaken and being pronounced dead at the scene. Would it really be worth it? Besides, what do you actually gain by not wearing your seatbelt in the first place?

A vaccine can be thought of as a kind of seatbelt for your immune system. In the same way that seatbelts and airbags can't prevent accidents from happening, they can significantly reduce your risk of serious injury in the event of a collision. In a similar way, vaccines can't actually keep you from coming into contact with dangerous viruses and bacteria, but they do offer a measurable and demonstrable level of protection in the event of an infection.

Sometimes I'll just casually mention that I assume Johnny Goodluck doesn't have any sort of insurance either. I'll admit that I'm playing the percentages here, because it's a fairly safe bet that most people have at least some basic form of insurance, you know, just in case something happens. This is a useful method of exposing an unconscious double standard, as long as you do it in a kindly and good natured way.

After all, if somebody's insured against accidental death or injury then it's a de facto admission that the unknown and the unpredictable is always just around the corner, and it's not always good. So then the question becomes, why bother to insure against unlikely events while taking a much bigger risk with your personal health?

Don't forget that a vaccination does something that no insurance policy can ever do, and that's to prevent a bad situation from escalating in the first place.

Any security expert will tell you that if you have limited resources, the best way to use them is to invest in preventive measures like locks and lights to protect your property, rather than buying an insurance policy designed to help cope with the aftermath of a break-in. As well as the seatbelt in a car, a vaccine can also be thought of as a good-quality door lock. It can't stop the undesirables outside from rattling the handle, but they'll soon pass on by and head for the house down the street with the open door.

Prevention is always better than cure, and vaccines are one of those rare and inexpensive measures that can actually prevent illness from developing, while medical treatment and health insurance can only help to deal with something once it's already happened.

In the final analysis, this kind of conversation always revolves around possibly the five most dangerous and irresponsible words in the entire English language: 'It won't happen to me.' But car crashes do happen sometimes, and if we accept the benefits of protection offered by seatbelts, airbags and insurance, then it makes a lot of sense to accept the similar benefits offered by vaccination, even if you're doing just fine without them . . . for now.

2

BUT VACCINES ARE FULL OF TOXIC CHEMICALS

I've encountered this general line of argument quite a lot over the years, and it's usually a pronouncement from those who follow a more 'enlightened' and environmentally conscious lifestyle, and are keen to tell everyone about their veganism and Bikram yoga classes. So when someone proudly displays their lifestyle credentials by informing me that vaccines contain toxic chemicals, the first thing I do is wholeheartedly agree with them.

All chemicals are toxic if they reach sufficient levels in the human body. This inescapable truth applies just as much to ubiquitous substances like table salt as it does to other, less benign-sounding chemicals like phosphorous, which is actually a naturally occurring element in its own right and also one of the necessary ingredients for DNA.

We'll not even get started on the hidden hazards of chemical compounds like caffeine in coffee and myristicin in nutmeg. That's for another book.

Persuading people to look at things differently nearly always stems from the judicious and consistent application of an undeniable truth – but often from a different perspective not previously considered. Yes, of course there are toxic chemicals in vaccines. As a matter of fact, it could not be otherwise.

Modern medicine simply would not exist were it not for the careful, studious and tightly controlled use of toxic compounds where there is a demonstrable medical benefit.

There's a good reason why the majority of medicines are only available from a licensed pharmacist, dispensing a prescription issued by a licensed doctor. The medical profession knows full well that drugs can be very dangerous indeed if misused, and the vast majority of the public knows it too. That's the simple reason these substances are not simply sold over the counter.

To put it another way, if somebody's aversion to the perceived toxic chemicals in modern pharmaceuticals is so great, would they be willing to quite literally stake their life on it? If they refuse vaccines, then logic suggests they should refuse antibiotics, painkillers, anaesthetics and chemotherapy too.

Naturally this line of reasoning can make some eco-warriors a little uncomfortable, as none of us likes to have our inconsistencies laid bare by a stranger. At this point I usually ask whether someone who is reluctant to vaccinate takes, or has taken, any other kind of prescription or even off-the-shelf medication. The answer to this question is nearly always yes, implying that the doubter believes there is something inherently different or sinister about vaccines in comparison to other classes of prescription medication. Often this might be because they have specific objections regarding the alleged toxic ingredients a vaccine could contain, with the two main offenders being mercury and formaldehyde.

'. . . But Vaccines Contain Mercury, and Mercury Is a Neurotoxin!'

This is the most common specific objection I hear relating to vaccination, at least among those who declare that they're unwilling to expose their bodies to toxic substances, especially while they're not even ill.

Mercury is toxic, there's no doubt about it, and so the decision to include minute amounts of it in some vaccines is an interesting one. Why would a pharmaceutical company include such a controversial element in their products, knowing full well that its use can be exploited by those advocating non-vaccination or offering alternatives? There must surely be a compelling reason for them to do so.

First of all, it's important to repeat that there's a minute amount of Thimerosal (a mercury derivative) present in some vaccines. Not all of them! The reason it's there is to act as an antiseptic and antifungal agent, preventing various microscopic nasties from multiplying inside the vaccination vial while it's in storage. So in fact, that tiny amount of mercury isn't really part of the active vaccine formula at all; it's more like a kind of medical preservative to stop the product from spoiling and creating a very real and dangerous risk to patient health. It's just a pity that the irony of these naturally occurring microorganisms causing a demonstrable health hazard is often lost on the ecological objector. Live cultures may be great in your farmers' market yoghurts, but they're a menace in medicines.

To put this whole mercury thing into some kind of perspective, you can expect to ingest roughly the same tiny amount of mercury in a single serving of canned tuna or sushi as you can through a routine vaccination (assuming the vaccine contains any mercury at all, because many do not). If you enjoy tuna or sushi (and other fish), then there's a pretty high chance that you're going to eat them a lot more than just once or twice in your life, thereby exposing yourself to mercury doses many times higher over your lifetime than you'll ever receive through vaccination.

There have been no observable adverse effects recorded in humans as a result of Thimerosal use. The same cannot be said for mercury poisoning caused by consuming large quantities of certain fish. Thankfully this is a rare occurrence, and there's no reason to stop eating your favourite fish dishes after reading this. However, it's a statistical fact that eating sushi is far more risky than getting vaccinated, at least as far as the danger from toxic mercury is concerned.

Both risks are very, very small, but the vaccine risk is still much, much smaller than the sushi danger.

The bottom line is that anyone who is worried about mercury exposure through vaccination should realise their mercury exposure levels are much higher via other sources, like eating tuna and lobster. They would have to conclude that either both eating fish and Thimerosal are an unacceptable health risk to the population at large, or that they're both likely to be relatively safe.

'. . . But What about Formaldehyde?'

Yes, formaldehyde is also toxic, and yes, you will also find tiny amounts of it in some vaccines. Just like mercury, formaldehyde is there for a scientifically valid reason and it has not been added simply to poison people.

Any readers who are thinking gruesome thoughts about Damien Hirst's infamous pickled cow are actually watching formaldehyde's vital function in vaccines played out before their eyes.

For those who aren't familiar with the story, controversial artist Damien Hirst rose to prominence some years ago when he displayed half a cow preserved in formaldehyde as part of an exhibition. His use of this particular chemical was by no means arbitrary and tells us a great deal about what's going on inside that tiny vaccine vial. Hirst could've suspended his demi-cow in water, or any number of other chemical compounds, but his biological exhibit would have either begun decomposing or even dissolved into its surrounding chemical bath.

Formaldehyde is useful in the way that it kills biological organisms while preserving their original molecular structure, just like it preserved the pickled cow.

What is a vaccine? Broadly speaking, it's often an inactive dose of whatever disease the patient is being vaccinated against, be it influenza, polio or hepatitis virus. Common sense tells us that we can't just inject someone with a live dose of HepB virus, as we'd simply be infecting them with the very disease we're intending to inoculate against.

In the laboratory, some vaccines start out as fully fledged, live and dangerous pathogens. Formaldehyde kills these microscopic organisms but leaves their molecular structure, their fingerprint if you like, intact. This is essential because it enables the human immune system to recognise the pathogen as a threat without being overwhelmed by it.

That's how some vaccines actually work. They're not really a syringe full of some mysterious chemical compound that somehow imbues immunity, but rather they can be thought of as a biological mugshot, distributed to the police force of our immune systems. Once the police force knows the villain's face, it's on patrol 24 hours a day just in case he shows up.

One of the lesser-known facts about using formaldehyde in vaccines is that it's actually removed once it's done its job and killed the active pathogen therein. Admittedly this is not a perfect process and a trace amount is left behind, even though it's no longer needed.

Anyone who knows a scientist will also know that scientists like to deal in specific numbers, whether they're really, really big numbers or unimaginably small ones. You can take it as read that if a scientist resorts to a non-specific word like 'trace', then the amounts they're talking about are very, very, very small indeed. So small that not even the nerdy guy in the white coat finds it necessary to specify exactly how small.

That's how much formaldehyde you can find in a vaccine: trace amounts.

Just like our example with the fish, formaldehyde can be found in much higher concentrations in many common fruits and vegetables. For example, the formaldehyde content of bananas and carrots is far higher than the trace amounts found in some vaccine formulae, and I've yet to meet a health food freak who'd never eat bananas and carrots because of formaldehyde.

Formaldehyde is in fact a naturally occurring substance which we consume in tiny amounts every day from most foods. The average apple contains roughly the same trace amount of formaldehyde as a single dose of vaccine, and as we all know, an apple a day keeps the doctor away.

Boiled down, objections to vaccinations on the grounds of toxicity can easily be addressed with just a little

knowledge of the subject. Vaccines do indeed contain minute amounts of excipients which are ubiquitous and virtually unavoidable in our everyday lives.

There's a perception problem with trace toxins in vaccines because they've gotten swept up in the same health and lifestyle concerns as food additives, being wrongly thought of in the same way. While many advocates of organic locally farmed food might forcefully argue that food additives are unnecessary, the same cannot be said for vaccines. Not only are the chemicals in vaccines vital to ensure their safe and effective use, but it's always worth remembering that while an apple is for an hour, a vaccine is for life. For those readers who might be thinking this argument is a false equivalency, because the toxic chemicals in apples and fish are ingested while vaccines are injected: Don't worry, we'll get to injected toxic chemicals later on.

The real question here is whether somebody who wants to live an unselfish, ethical lifestyle is willing to bear the statistically miniscule personal risks associated with vaccination, in order to secure the massive individual and collective benefits derived from the near elimination of diseases such as polio and smallpox.

3

VACCINES MAKE ME FEEL SICK

Okay, so I'm kind of paraphrasing a bit here, but that's a reasonable summing-up of probably the single most surprising objection I've heard over the years. Although what I'm much more likely to hear from an adult is that their child felt poorly after a vaccination, or a friend's child felt poorly, or they'd heard about some kid at the same school who was ill, or maybe it was the same town, but a different school . . .

Following my tried and tested rules of engagement, I try not to get into the nerdy and technical stuff surrounding normal reactions to vaccines, but it never does any harm to acknowledge the truth that some people can feel a bit poorly after getting a shot. There are a couple of very good reasons for coming clean when something like an unpleasant side effect comes up.

The first reason is that it makes me sound a lot less like a salesman, because I'm not actually selling anything. The brutal truth is that it makes no difference to me whether the person I'm talking to gets vaccinated or not, but it could make a big difference to them and their loved ones.

The second reason is that there's very little in life that's an absolute gain with no downside anywhere along the line. Cars cause pollution, machines threaten jobs and sometimes vaccines can make people feel a little poorly for a while.

Believe it or not, that's actually a good thing.

As we've already discussed, a vaccine is an inactive dose of a real-life, bona fide virus or bacteria that we use to train our immune systems to recognise as a threat. Post-vaccination symptoms like headaches, temperature and a slight fever are in fact indications that the body's immune system is working overtime as it learns how to combat a previously unknown pathogen.

Obviously there's little point trying to explain that to a toddler, but then it won't be the toddler who books his next doctor's appointment. The best I can do is to help a worried parent understand that what's happening to their uniquely special and precociously talented princess is completely normal, and is in fact a natural part of the immunisation process.

I've always found it helpful to remind concerned parents that the unpleasant side effects of vaccines are short-lived and that the virus or bacteria it contains is a minute fraction of its normal strength. The mechanism of action and individual immune responses can vary between different vaccines for complex biological reasons that I won't bore you with here. That's why someone may receive one vaccination and feel fine, while they may feel a bit under the weather after the next one. There's really no telling who might feel a bit ill and who won't before they're vaccinated, but severe adverse reactions are extremely rare indeed.

The way I put the point across is something like this: If you've ever been a bit poorly after a vaccination, just try to imagine feeling ten times worse! That's what you'd feel like if you were ever unlucky enough to contract the full-blown disease, and it would last an awful lot longer than 24 hours, too.

Would you really want to risk that for yourself, let alone your children?

In the end, vaccination is pretty much the same kind of cost/benefit choice that dominates so much of our lives, whether we realise it or not. It's choosing between maybe feeling ill for a day or two, and running a greater risk of contracting a serious and even life-threatening condition. It's a small inconvenience in exchange for a near lifetime of protection from nasty diseases, and hopefully it's an easy choice.

If getting your shots isn't a no-brainer now, it certainly becomes one if you've ever been afflicted with a vaccine-preventable infection. Take the example of my friend Billy; no, that's not his real name, although I can assure you that both he and his story are real enough.

Billy's a pretty smart and ordinary kind of guy. Like most people, he's not a scientist, although he doesn't wear a tin-foil hat either. Neither does he loudly insist on eating only wholesome free-range and happy vegetables packaged in rustic-looking crates by a multinational company. However, for reasons I've never really asked him about, Billy missed some of his childhood vaccinations.

In common with a lot of generally healthy people, Billy thought of those missing shots as a hassle; and besides, he'd had a few travel vaccinations before going on holiday some years before and they'd made him feel pretty rough for a while. Then one day, not so very long ago, Billy caught chickenpox as an adult.

Anyone who has contracted that infectious disease while they were still a child will immediately tell you about the fevers and red spots. Above all, though, there was the constant, unbearable itching that the ubiquitous

chamomile lotion did pretty much nothing to soothe. They'll tell you how their parents constantly monitored and admonished them for scratching, and how the only upside was missing a few days of school.

Billy wasn't lucky enough to catch chickenpox when he was a child, meaning that when the disease struck, he felt its full fury as an adult.

As an interesting aside, I'm often asked why certain 'childhood' diseases like measles and chickenpox seem to strike adults with far greater ferocity. As this is a book that tries to avoid long and nerdy scientific discussions, I'll simply say that nobody really knows for sure. There are plenty of ideas and theories kicking around at the moment, but even in the twenty-first century, our scientific understanding of something that we anecdotally know to be true is not nearly as complete as people often think. That surprises a lot of folk, but it's a handy way to demonstrate a little humility before getting back to talking about poor old Billy.

Having contracted chickenpox later in life, Billy was laid up for three weeks with a dangerously high fever, chills, sweats and intermittent delirium just for good measure. He was in constant pain with more than three hundred blisters covering his body. In the end, he became so ill that he was hospitalised for several days.

After his ordeal, Billy confessed how foolish he felt about putting himself through all that torment just because he didn't want the inconvenience of feeling a little bit ill for a day or so. In the light of his chickenpox experience, he knew that maybe feeling a bit rough for 24 hours was more than a fair exchange for a near lifetime of protection.

When he viewed that whole situation in the light of his recent illness, Billy realised just how irrational his thinking had been when it came to vaccines.

You'll be happy to hear that Billy's made a full recovery and certainly won't be making the mistake of not keeping his shots up-to-date. In short, if you think vaccines make you feel sick, you'd better believe that the infections they protect you from will make you feel way sicker.

4

VACCINATIONS CAUSE AUTISM

No they don't. Seriously, they really don't.

I simply can't say it any clearer than that.

I wish I did know what causes autism, but everything we presently know points to the fact that vaccinations do not.

This is especially frustrating because, despite the spurious link between autism and vaccines being a notorious and well-documented case of fraudulent science, it still somehow manages to linger in the odd corner here and there. Kind of like a bad smell that you just can't get rid of.

It's also saddening because it really demonstrates how wrong information can take root in the public consciousness very quickly, despite having more data readily at our fingertips than at any time in human history.

Believe it or not, the whole vaccines and autism thing got started from a single, very flawed and deliberately distorted study that caught the news media's attention back in 1998.

That year, a scientist called Andrew Wakefield published a report in the *Lancet* claiming there could be a link between the MMR vaccine and autism in young children.

Not surprisingly, the general press soon got hold of the story, and hey presto, a major health scare began brewing. Worried parents stopped vaccinating their children in large numbers almost overnight. I mean, why wouldn't they? After all, Wakefield's research paper had been published in a respected medical journal and not some obscure underground, pseudoscientific publication.

Naturally, the vaccine manufacturers and medical professionals did their best to put out the flames, but the damage had already been done. That single health scare led to such a significant downturn in vaccination rates that dangerous diseases like measles started to reappear in places they'd not been seen in serious numbers for several years.

At the time the study was published, there were a great many number of professionals who questioned Wakefield's data, methods and conclusions, but they were usually drowned out by the growing hysteria or, even worse, dismissed as propagandists for the big pharmaceutical companies.

Despite the story eventually disappearing from the front pages, the damage had already been done and everyone 'knew' that there were some serious health concerns surrounding the MMR vaccine.

Although many qualified professionals were suspicious early on, it wasn't until 2010 that the *Lancet* finally published a full retraction of Wakefield's work, no less than twelve years after his initial discredited paper was released.

By the way, so grievous was Wakefield's deception that he was eventually struck off by the British Medical Association and prevented from practising medicine.

Of course, Dr Wakefield's spectacular fall and subsequent disgrace were dutifully reported in the press, but stories about defrocked doctors just don't travel as far and wide as lurid tales of children in danger and corporate deception.

However, every cloud has a silver lining and it does make me happy to hear the autism objection a lot less these days. If it does come up, I just give a very quick, potted history of the Wakefield case and suggest that people Google it for themselves. They can do it right there and then on their smartphones if they're not willing to take my word for it. This tends to allay a lot of concerns, and I usually hear something like, 'Gosh, I had no idea it was all a scam' or 'Do you mean I've been worrying all this time for nothing?'

Yes, I'm afraid you have.

It's true that although the autism objection is deservedly dying, that medical drama has created a small but rather difficult-to-reach group who simply refuse to believe that the MMR vaccine is safe. I like to call them the moon children, but I'm only half mocking them.

I call them the moon children because these are the conspiracy theorists. These are the people who choose to believe that the Wakefield report was true, and that various powerful but poorly defined 'interests' have somehow managed to exert enough pressure on this discredited doctor to have his research consigned to the dustbin of history, all to keep the autism–vaccine relationship hushed up.

This is why I call them the moon children, because they're the same sort of selectively sceptical types who also claim that the moon landings never really happened. It's the kind of glib statement that people sometimes make without really thinking it through – although that's exactly what I invite a moon child to do if I find myself face to face with one.

So let's think this whole moon landing thing through for a minute.

To date, the theories of how NASA might have faked the landings are scarcely convincing, but most damning is the fact that no one gives a credible account of how they could've possibly managed to keep it all a big secret for so long. In order for such a vast conspiracy to be viable, all the Apollo astronauts would need to be involved, as would a significant number of senior staff at NASA, plus a whole slew of senior government sources. Not only that, but they would have to hoodwink their own middle management and staff as well as the public.

It doesn't stop there, though, as they'd also have to maintain a strict code of silence for the rest of their lives and even beyond the grave to ensure such an elaborate charade could remain undiscovered. Given how we know that large organisations and governments struggle constantly with leaks and rumours, we're forced to ask ourselves this simple question: Just how likely it is that such a massive deception could be maintained in perpetuity?

Just thinking about the logistics of keeping something like a fake moon landing quiet is enough to cast serious doubt on any such radical conspiracy theory. The people involved would need to be constantly monitored forever, just in case they let something slip. Such an undertaking would require a massive allocation of resources and manpower, and the irony is that the huge team charged with ensuring absolute secrecy would itself become a serious security headache. Anyone with any experience of keeping secrets will tell you that the risk of leakage only increases with each new person who's brought into the loop.

In addition, if such an enormous and elaborate hoax like the moon landings were ever true, it would create a huge and irresistible incentive for some insider to expose it. Money, fame, political power and probably a combination of all three would eventually prove too much of a temptation for someone who knew he could demonstrate the deception for sure. We'll not even start with the Soviets, who would've had a very clear motivation to expose the landings as fake, since losing the space race was a huge blow to their image as a world superpower.

The same logic applies just as much to the autism scare as it does to the moon landings. If you believe that the link between vaccines and autism is real, then you also have to believe that Andrew Wakefield's research was basically sound. Added to that, you have to accept that the medical professionals and academics who discredited his research are part of some nefarious conspiracy to protect the reputations of big pharmaceutical companies. If you can get over that hurdle, you then have to continue believing that all other medical professionals, academics and researchers in the vaccinology field all over the world are either similarly corrupted or tacitly agree with the revised (and supposedly wrong) findings. While it's true that big drug companies can make an awful lot of money, it's doubtful that even their pockets are deep enough to buy off a worldwide and, let's be honest about it, well-paid profession in its entirety.

If it's a question of money and power, then just imagine the fame, fortune and prestige that would be showered on the scientist who could conclusively prove a causal link between various vaccines and incidences of autism in children. If I ever found a chemical compound directly

and irrefutably linked to autism, I can guarantee I'd be hailed as a hero and a benefactor of the human race, and probably win a Nobel Prize. I'd be set up for life and essentially able to do whatever I wanted for the rest of my days. So what would be my incentive for keeping quiet?

As a scientist, if I believed for a moment that the MMR vaccine was really linked to autism, I'd be shouting about it in every journal and to every journalist I could find, as would most other scientists around the world – firstly because it would be the right thing to do, and secondly because it would make me a household name overnight.

The simple and reassuringly dull truth is that vaccines don't cause autism, so I guess I'm going to have to look somewhere else for my big scientific breakthrough.

When dealing with the moon children, your best chance of success lies in illustrating just how unlikely such a large-scale conspiracy would actually be, rather than just insisting that vaccines are great; it can be quite effective. Once or twice I've even heard the words, 'Huh . . . I hadn't thought about it like that.'

5

VACCINES ARE NOT FOR ME

This argument usually comes in two flavours. Some people believe that vaccines are really meant for old and sickly people, and since they are fit and healthy individuals vaccines are not really applicable to them. Admittedly, many of us like to think we're invincible, especially while we're in the earlier part of our lives. After all, why shouldn't we think that way? We tend to shrug off minor illnesses and injuries with relative ease while we're young . . . and ironically, that's where the problem lies.

The innate physical advantage of youths and young adults might go some way to explaining why study after study has shown that most of us are pretty useless at appreciating risks and calculating the odds concerning our future well-being. Just talk to any smoker and he'll likely tell you that he already knows that smoking greatly increases the risk of lung cancer. What he's far less likely to know, or want to know, is just how much that risk increases. To put it in simple mathematical terms, if you smoke, you're 23 times more likely to develop that life-changing and often life-shortening condition than if you don't.

The same goes for riding a motorcycle. Any biker and most car drivers will know that riding a motorbike is a lot riskier than driving a car. Just like the smoker, though, most road users don't realise that the average motorcyclist is no less than 35 times more likely to die on the roads than a car driver.

Of course, just knowing those facts won't make a smoker quit on the spot or persuade a biker to hang up his helmet. Human behaviour is a great deal more complex and irrational than a simple but mathematically valid risk assessment, which is why I seldom bother pointing out the downsides of non-vaccination in terms of numbers and percentages.

When it comes to illness and vaccinations, a strong, healthy immune system can actually lull the young into a false sense of security. It's easy to think that because we can easily shrug off a cold or some minor injury we can deal with anything more serious that comes along. Chances are, the more fit and healthy a young person is, the more likely he is to adopt that kind of invincible out-look; whereas a younger person who is asthmatic is much more likely to get that flu jab, along with all the others. Nobody needs a doctorate or a medical degree to know that the asthmatic guy has already suffered disproportion-ately at the hands of respiratory diseases like colds and flu, and thus he has learned about his own vulnerabilities far earlier than the super-fit sports jock who poses for pictures before dancing over the touchline.

However, the sad and unsettling truth is that young people do sometimes suffer from life-changing and poten-tially lethal diseases such as hepatitis B or meningitis C, both of which can be controlled by vaccination.

However, it's not always invincible millennials whom I hear this kind of argument from.

Older people don't tend to think they're invincible because they know better, but in demographic terms, many of them are kids from the '60s and '70s, which makes them either the drivers behind or the descendants of the hippy movement that once was. In common with the millennials, they share an instinctive distrust of large-scale medical interventions like vaccination. They often believe that vaccinations are 'not for them' since these are either useless, counterproductive or maybe even some kind of pernicious state slavery project, depending on how much patchouli they consumed back in the day. These are the spiritual children of the flower power era, the ones who always look for a 'natural' remedy for their ills . . . whatever that means. Their objection to vaccination is usually based around a general desire for a more harmonious, natural and collective life experience. This demographic tends to reject both the starchy conservatism of the post-war era as well as the rampant consumerism of the '80s and '90s.

In their eyes, state-sponsored vaccination is viewed with the same degree of suspicion as modern farming practices like chemical fertilisers, pesticides and even genetically modified crops. This attitude forms part of an overarching umbrella of belief that most of these modern techniques are designed purely to profiteer at the expense of human health, harmony and general well-being.

There's clearly quite a lot of overlap between the way the younger and older groups arrive at the not-for-me attitude, which is why they both show an interest in the way I talk to them about vaccinations.

I've already mentioned some facts, figures and statistics in this chapter, and although they're perfectly correct and verifiable, none of them is likely to make any particular impression on this variety of doubter.

What is very persuasive is to harness the invisible bond that ties these two seemingly disparate groups together, and that bond is best described as an urgent and over-riding sense of social responsibility.

Whether you're talking to a 'woke' Williamsburg activist or a wizened, sundried organic farmer, the one thing they will both relate to is an appeal to the little guy, the victim, the marginalised and disenfranchised, and our civic duty to protect them.

The young and invincible might not be too worried about getting sick themselves, but rather than hitting them with facts and figures, I just explain that getting vaccinated helps everyone, especially the meekest and most vulnerable individuals in society. An appeal to the social warrior within them immediately transforms the decision over whether or not to get those shots from a personal lifestyle choice into an issue of social and moral responsibility. Indeed, when someone says 'not for me', they're failing to recognise that vaccinations aren't just for their own personal benefit; they're for everybody else's too. It's always worth reminding them that vaccinations work by training the body's existing immune system to recognise a potential threat early before it can take hold, and this is especially effective in healthy young people who already have a good immune system. People just like them. Vaccines are far less effective for the very old and those with a compromised immune system, and so those individuals who need it most often can't benefit from vaccination. It's really up to the healthiest in society to help protect the most vulnerable groups like infants, the elderly, cancer sufferers, transplant recipients and HIV patients by getting immunised. Vaccinated healthy people protecting the more vulnerable groups is an essential mechanism of vaccine efficacy, often referred to as 'herd immunity'.

Herd immunity basically works by surrounding vulnerable individuals with healthy people who've been immunised, creating a sort of human barrier to prevent the spread of infection.

This kind of socially responsible outlook is also helpful for persuading older people that vaccinations aren't merely a deeply personal choice that only affects them. In fact, getting immunised is the morally responsible thing to do as it protects both them and those around them, especially those who can't develop a good, strong immune memory of their own.

When presented in this way, people who have previously stated that they don't do vaccinations can see that it's really a socially responsible act, just like using public transport, recycling or voting. Unvaccinated people who make the change aren't just helping themselves, but they're helping to make the world around them a safer and healthier place too.

The truth is that vaccines are indeed 'not for you'; they're for everyone else too. Making that connection between getting vaccinated and the social responsibility of protecting 'the herd' is very often an effective way of discussing the benefits of vaccines with someone who shows signs of a strong personal morality.

6

THEY ONLY WORK HALF THE TIME

I don't come across this idea very often, but when I do I know I'm hearing a classic case of reaching a wrong conclusion derived from partial information.

It's certainly true that some vaccines are more efficient than others, although professionals like myself are working hard to increase the effectiveness of all vaccinations. At the moment the picture is pretty mixed, even though we're always making progress. Just by way of an example, the measles part of the MMR jab can reach above 99 per cent efficacy, while scientists are still searching for a safe and reliable malaria vaccine, a version of which has shown an efficacy range of 30–50 per cent in recent trials.

As this book is more of a conversation than a scientific textbook, I'm not going to stray too much into the weeds of why some vaccines offer better protection than others; but if I'm ever pressed on the issue, I usually point out that it's a combination of factors. These include the virulence of the pathogen concerned, vaccine components, mechanism of action, methods of vaccine production and each individual's unique physiology.

Although it's tempting to use analogies like crash helmets or so-called bulletproof vests to illustrate the point that some protection is better than none, those ideas miss out a crucial and often understated difference between mechanical safety measures and biological defences like vaccinations.

That's right; we're back to the 'herd immunity' idea once again. If everyone rides a motorbike, and crash helmet technology only protects, let's say, half the population in the event of a collision, then you have a 50 per cent chance of emerging unscathed if something should go wrong. The fact that the guy next to me at the traffic lights enjoys the same level of protection that I do doesn't actually help me in the slightest. His level of protection is neither affected by nor dependent on mine, and vice versa. However, vaccines don't work in the same way, especially in large and tightly packed human populations. Voting is usually a better analogy. While it's true that a single vote can't change an election, it's important that as many people as possible vote in order to deliver a representative result. If nobody shows up, then the political system is poorly protected against less savoury and more insidious elements and the welfare of vulnerable minorities can be overridden.

The inescapable mathematical truth is that where vaccines are less efficient, it's actually more of a reason to get your shots and not less. If a certain vaccine is in the 50–60 per cent efficiency range, then that means a large percentage of the population will have to be immunized before the 'herd immunity' threshold is reached. In that scenario, there will be plenty of people walking around who aren't protected, unwittingly lowering the herd's immunity level even though they've done the right thing and had their shots.

Deliberately leaving yourself unprotected when a vaccine is less effective is even more irresponsible than doing it where effectiveness is high. As we've already discussed, vaccines aren't just for you, they're for everyone around you as well. The less efficient a vaccine is, the more that old adage holds true.

We all need to remember that it's not just ourselves who are affected by our vaccination decisions; we help everyone by getting a shot. That makes each of us either part of the problem or part of the solution.

7

I'D NEVER INJECT MAN-MADE CHEMICALS

It's easy to conflate this fairly common argument with our earlier chapter on toxic chemicals. I'll be the first to admit that while there are some similarities, they do also differ in one very important way.

The toxic-chemicals discussion deals with concerns about the known toxicity of certain specific chemicals, especially mercury and formaldehyde, because everybody knows those substances are dangerous in sufficient amounts. However, this is much more of a general argument that I've come across more than a couple of times during my career.

Boiled down to its basics, this kind of objection is usually rooted in a general misunderstanding of chemistry and biology, which leads to a natural suspicion regarding all manner of mysterious and scary-sounding chemical compounds. There are several different shades of this argument, but they all revolve around an objection to the idea of injecting mysterious chemicals with long and unpronounceable names.

When you stop to think about it for a second, nearly everything, including us, can be described as a bunch of different chemicals interacting; it's just that we don't usually talk about the world in those terms because we don't need to.

Probably the best way to highlight this intuitive mistrust of chemical compounds with unfamiliar names is to draw attention to the subject of dihydrogen monoxide, or DHMO as it's more commonly called. This potentially lethal chemical is all around us, in the atmosphere, in the water supply and even in our very homes. Despite numerous campaigns to raise awareness of just how prevalent this often corrosive chemical can be in the environment, the general public remains woefully ill-informed about the potential hazards associated with DHMO. Don't take my word for it, as there are entire websites dedicated to the study and analysis of DHMO dangers, as well as useful fact sheets to help worried parents and factory employees understand the potential hazards of this ubiquitous chemical compound. It's every citizen's civic duty to be at least minimally informed of the dangers surrounding dihydrogen monoxide, but still the mystique and lack of public awareness surrounding DHMO persists.

If you haven't figured it out already, DHMO was a hoax popularised via the Internet by using a more scary-sounding name for H_2O, commonly known as water. Yeah, that's right, dihydrogen monoxide is just plain old water, described using more technical but nonetheless correct language.

Incidentally, there really are some great spoof websites and safety guides for DHMO if you'd like to go and take a look at them – firstly because they're generally amusing, and secondly because they do illustrate a very important point about perception and reality not always matching up.

Just because something has a long and mysterious-sounding chemical description, that doesn't necessarily mean it's toxic, corrosive or otherwise untrustworthy. Sometimes these generalised attitudes can manifest themselves in some novel ideas as to how to judge the safety or wholesomeness of any given chemical or vaccine. I've even come across someone who confidently asserted that 'if you wouldn't eat something then you shouldn't inject it either'. Admittedly this was on Twitter, so I don't know how seriously we should take it, but it did highlight a pervasive anxiety surrounding syringes and what's inside them.

The hard scientific truth is that trying to administer all pharmaceuticals orally makes about as much medical sense as injecting guacamole or barbecue sauce straight into the bloodstream. As an experienced biochemist, I would urge readers never to attempt anything so foolish and irresponsible.

At the end of the day, food is food and injected medicines are injected medicines. They are two different things that both perform vital functions for the protection and main-tenance of health. I can't believe I actually have to write this simple observation down, but it does at least give me a chance to talk about injections in a little more detail. Not too much detail, as I know a lot of people don't like needles. However, it's good to bring up the subject here because I know that syringes are one of those frightening medical items that can encourage the perception that if some vaccine or medicine needs to be injected, it must somehow be more serious or extreme than simply pop-ping a pill.

Believe me, if there were a safe and effective alterna-tive to injections then I promise you we'd already be using it. Just imagine how rich the guy who finally fig-ures out how to eliminate needles is going to be! In any case, when vaccines can be taken orally (polio, rotavirus, typhoid, etc.) it has become the norm to administer via this delivery route. Why go to the bother and expense of injecting something when you don't have to?

The bottom line is that we use injections because there's no viable alternative that is as safe and effective. If it's any reassurance, I'm a professional and I don't like needles either, but I'll accept an injection because I know it's the only way to receive certain treatments, including most vaccinations.

However, despite some strong objections to injections generally, and injecting man-made substances specifically, I can't help but notice how often the same people who refuse to inject manufactured chemicals do not appear to have any qualms about the growing trend of getting inked up with various permanent tattoos. On a daily basis, thousands of people willingly have themselves injected with chemicals like disazopyrazolone, ferrocyanides, lead carbonate and many other chemicals commonly found in tattoo inks. Despite the chemical reality of tattooing, permanent body art is seldom viewed as a nefarious health risk.

Although I have no reason to think they're doing it consciously, tattooed objectors to man-made chemicals are in fact displaying a pretty hardcore double standard.

Tattoos aren't just one injection; they're literally hundreds of small puncture wounds, each of which deposits, you've guessed it, man-made and mass-produced chemicals under the skin. What's perhaps even more surprising is that a significant number of tattoo inks started life as chemicals associated with industrial processes like painting cars, and they still contain some of the same ingredients.

This double standard becomes even more nonsensical when you consider that tattoos can be very expensive and yet deliver no discernible health benefits. A tattoo can't help to protect you from a virulent disease in the same way that a vaccine can, and in fact the statistical evidence shows that getting a tattoo is in fact a greater risk to your health than getting a vaccine.

For starters, tattoo inks are far less scrupulously tested for safety than vaccines. That's not to say that tattoo inks are inherently unsafe, but it's an inescapable fact that they do contain far higher amounts of potentially harmful substances than vaccines. There's also the ever-increasing risk of fakery to contend with in the body-art market. We've all seen those TV programmes where cheap knockoffs of everything from designer handbags to electrical equipment are flooding the Internet, and tattoo inks are no exception. Even though your local tattoo guy is probably a straight shooter, there's always the risk that he might be taken in by a good-quality fake of an established

and legitimate brand of ink. If that's the case, then God only knows what might end up circulating through your system if he inadvertently uses something that's not designed for the purpose.

Because pharmaceuticals, including vaccines, are so tightly regulated and exhaustively tested, there's very little chance of any potentially harmful forgeries making their way onto the market and into the doctor's office.

Now I'm not here to disparage tattoo artists, and I know that the overwhelming majority of them take safety and hygiene very seriously. Unfortunately, though, this is one occasion where you just can't argue with the numbers. It's a mathematical fact that you're much more likely to experience some kind of allergic reaction or skin infection after getting tattooed than you are after getting your vaccines. There are probably a number of different factors that contribute to this increased risk, including poor hygiene practices, lack of consistent testing and rogue, irresponsible tattooists. You can probably minimise a lot of those factors by going to an established and reputable tattoo artist if that's your thing, but your statistical risk is still higher than going to your doctor to get a vaccine.

This is probably the biggest irony I encounter when I'm talking to people who won't inject man-made vaccines but have zero objections to tattoos. By the way, I have no personal objection to tattoos or tattoo artists, and I believe the practice to be generally safe. But I suppose that at the end of the day it boils down to one simple argument: If you have a strong objection to vaccines based on a refusal to inject 'unnatural' man-made chemicals into humans, then you should probably be just as vocal (if not more) against the practice of tattooing.

8

I'VE NEVER HEARD OF THESE DISEASES

It probably won't surprise you to learn that it has been someone quite young saying this in almost every case.

In contrast, talk to anyone past retirement age and there's a high chance they'll not only have heard of these diseases, but they'll be able to name at least one kid they went to school with who was permanently affected by polio; or they'll tell you that they themselves were once quarantined after contracting the measles virus. These once common and feared diseases are still a part of living memory, and it's a testament to how far and how fast we've progressed that they're now so rare that fewer and fewer young people have even heard of them.

Indeed, the World Health Organisation reports that in 2016 the top two causes of death globally were heart disease and strokes. That statistic probably isn't surprising to anyone living in the industrialised world, but nevertheless it marks a radical shift in humanity's development. Before the advent of mass immunisation programmes, the most common causes of death were respiratory infections or diarrhoeal disease caused by viruses and bacteria. It's no coincidence that these are still the two most common causes of death in the developing world, where immunisation programmes can be either non-existent or patchy at best.

When you consider how the development of vaccines during the late nineteenth and twentieth centuries has literally changed the course of human history, the

importance of vaccinations simply cannot be overstated. Indeed, along with the development of antibiotics and improved sanitation, it would be very reasonable to claim that vaccination is one of the most important medical breakthroughs the human race has ever made. Although it's very difficult to prove a non-event such as somebody not dying, it's no exaggeration to say that there are literally millions, if not billions, of people alive today as the result of a simple vaccination jab. It's as close to miraculous as the medical profession is ever likely to get.

While all of this is a testament to the effectiveness and vital importance of vaccinations, it can also create a few unexpected challenges, especially among younger people. If familiarity doesn't exactly breed contempt, then it certainly does give rise to widespread and unconscious complacency. To put it bluntly, there are far too many people walking around who mistakenly believe that diseases like diphtheria and whooping cough fever somehow belong only in the past. In reality, with a couple of notable exceptions like smallpox, these pathogens are all still very much alive and out there in the world, just waiting to exploit any opportunity to spread and multiply.

Because our parents and grandparents grew up at a time when life was more precarious in so many different ways, they didn't really need much persuading to sign up for vaccinations. While the average person couldn't do very much about the constant menace of the Cold War, public vaccination programmes did at least give them the opportunity to protect themselves against the scourge of diseases that could strike anyone at any time. In fact, none other than President Franklin Delano Roosevelt struggled with the crippling after-effects of polio even while directing the United States through the Second World War. That's how prevalent these diseases were in the earlier part of the twentieth century. Nobody was safe, not even the president of the world's one and only hyper-power.

Today it's very easy to take some kind of moral or ethical stand against vaccinations when you have no skin in the game, but I often wonder how long those high-minded principles might survive if there were a polio or measles outbreak in the objector's own neighbourhood. Not so very long I reckon.

As is usual in my line of work, although everything I've said up to this point is true, that doesn't always help to convince someone who's young and with little interest in the past that vaccination is good both for them and for those around them.

Naturally as individuals and as a society, we tend to focus on the most immediate and serious problems we can see in front of us, and that's reflected in the major health worries for the developed world, which are currently heart disease and cancer. This is an idea I sometimes explore if someone believes that vaccines are only for 'old-time' diseases that we just don't see any more. Just ask yourself this simple question: If somebody perfected a vaccine against cancer or heart disease tomorrow, would you use it? I certainly would, and so would the vast majority of other people too. Just imagine how euphoric you'd feel if some egghead in a white coat came up with a simple injection that would give you a better than 90 per cent chance of being cancer-free throughout your entire life. Now imagine some teenager from the year 2085 telling you they've never heard of cancer and they don't know what the big deal is. Wouldn't you just want to slap him? Wouldn't you want to describe the suffering of your friends and relatives in graphic detail before physically dragging him to the doctor to get a shot?

Well, if you've never heard of *Haemophilus influenzae* or *Bordetella pertussis* then go and take a good look in the mirror, because you could easily be that kid from 2085, just dressed in different clothes and worrying about health problems that the kids from the future probably won't think are important. At least now you know what some older people are really thinking if you talk like that about diseases that once ravaged their friends and family.

Just as an interesting aside, there's actually a great deal of hard work and research being conducted into anti-cancer vaccines right now. Although we're still some way from mass immunisation, such an idea is no longer an impossible pipedream. Although it won't happen next week, or even next year, research results suggest that such a day may well arrive in the not too distant future.

Thomas Jefferson is often misquoted as having said that eternal vigilance is the price of freedom, and although that idea was expressed about politics, the same is equally true of dangerous viruses and bacteria. Today we live in a very safe and settled time biologically speaking, but that's only because those who went before us developed and effectively used the tools required to banish these once common killers to the museum and the textbook. However, the fact that we've prevailed in battle doesn't mean that the war is won and those diseases from our past have been eradicated entirely. The vast majority of them are still out there somewhere, and if we give them half a chance they'll creep back into our midst and strike right in the heart of our densely packed and highly vulnerable societies.

When you consider the unsettling idea that we might be approaching a post-antibiotic era, it makes a lot of sense to use every other tool we have, including vaccines, to ensure that the diseases we now have under control never have the chance to flourish again. The cost of complacency could be very, very high indeed.

9

IT'S JUST A BIG PHARMA SCAM

This is an attitude that has come up fairly regularly throughout my career, and it always leads to an interesting conversation because this is a specific idea springing from a broader philosophy regarding business generally, and big business in particular.

So let's start by just clearing the air and stating the obvious. In some respects, a big pharmaceutical company is no different from the coffee house across the road.

Both businesses exist in order to make money, and they can't continue to exist unless they do make money. Naturally, both businesses want to try and make as much money as they possibly can with the resources at their disposal. Both businesses aim to offer something that customers either want or need in exchange for monetary compensation, whether that something is a hot cup of coffee or a life-saving drug. If businesses don't produce something that's either useful or desirable, then they don't make money and they cease to exist. It's really that brutally, clinically simple.

Nowhere is this more true than in the world of commercial pharmaceuticals. Forget those loudmouthed money boys with their gaudy pinstripe suits, and believe me when I tell you that pharmaceuticals is one of the most ferociously competitive and merciless fields of business ever created. The end result of all that hard work and tough competition is hopefully a big profit for the drugs company and a life-changing product for the consumer.

In the real world, most of us don't actually work for a large multinational company like a pharmaceutical giant, so their activities and motives can often seem distant and opaque to someone on the outside. It's easy to lose sight of the individual benefits we enjoy as consumers when we're staring up at some towering fortress of steel and glass, wondering what on earth could possibly be going on in all those offices.

It's this kind of attitude that leads some people to take the view that they're not going to fork over their hard-earned money to some faceless corporation that is already extremely wealthy. Just like in the previous chapter, this sort of argument is sometimes a mask to hide a different concern such as vaccine safety or effectiveness. On the other hand, sometimes it really is all about the money, and about somebody not wanting to hand any of their finite budget over to some sprawling organisation that clearly doesn't need it anyway.

When I come across this kind of objection I tend to just ask a single, simple question: Would you accept the vaccine if it were given to you free of charge? If the answer is no, then we're likely to circle back to one of the other arguments discussed in this book. If the answer is yes, then I have some excellent news for the vaccine dissenter, because most of them are already free! How awesome is that?

If you're a child or a high-risk adult in the developed world, the chances are that you won't have to pay a penny for a lifetime of immunity from some very dangerous diseases, so a vaccination is likely to be one of the best deals you're ever going to get. You won't have to put your hand in your pocket at all to benefit from a near lifetime of protection. That should sound like a pretty good arrangement to anyone concerned with financial disincentives.

Of course, the true reason that people aren't usually asked to pay up front for vaccines is that the economic case for providing them free at the point of delivery is overwhelming and incontestable.

Healthcare costs and lost economic output could quickly cripple even a wealthy nation in the event of a preventable epidemic, and that's without considering the human suffering and social alienation that would arise from such a tragedy.

However, it's this same hard-nosed economic reality which has encouraged some people I've met to refuse vaccines on the grounds that pharmaceutical companies are driven primarily by making a profit, while helping humanity is only a secondary consideration.

As I said earlier, a pharmaceutical company wants and needs to make money just as much as a coffee shop does, but the fact that there's so much money to be made in developing useful drugs and vaccines ensures that competition is fierce and standards are high.

If the coffee shop makes a mistake and sells a stale pastry, then a simple apology and a fresh replacement can usually solve the problem.

There is no such mercy in the pharmaceutical sector. If a vaccine manufacturer makes a mistake, then the results can be nothing short of catastrophic. To begin with, there are the direct implications for the consumer who receives an ineffective or impure vaccine. Fortunately, serious health risks associated with substandard pharmaceutical products are extremely rare, firstly because of the tightly regulated drugs market, and secondly because manufacturers know that one single slip-up could mean the end of their entire operation.

Product recalls are extremely expensive and hugely damaging to any brand's standing and public image. Consumer confidence falls, leading to a further drop-off in revenues, which in turn affects stock prices and potentially leads to an irreversible downward spiral. That's before mentioning the added damage that could be inflicted by competitors just waiting in the wings to scoop up that lucrative government contract. When it comes to mass vaccinations, you can probably multiply that level of scrutiny and competition by a factor of ten.

Governments are acutely aware of public perception when it comes to taxpayers' money, and so they're always looking for the cheapest, safest and most efficient vaccines available. If a rival company can produce a product that's better and cheaper than the current supplier, then the state will not want to face the political fallout of continuing to supply a more expensive and less efficient vaccine.

In other words, despite some of the more lurid and dubiously sourced stories circulating online, drug companies know that there's nothing to be gained by producing a damaging or poor-quality product. In fact, all the economic and political incentives push the pharmaceutical industry to produce the best products that it's humanly possible to manufacture. This is what economists refer to as the discipline of the market, and you can believe me when I tell you it really works when it comes to vaccines.

While it's demonstrably true that corporations can only survive by doing something useful for their customers, it's also true that one or two of them might occasionally behave unethically in the pursuit of market share and profit. This is a much greater danger in other sectors like finance or insurance, since by comparison big pharma is much more closely monitored by governments, regulators, the press and consumer advocates so that any kind of shady or unethical behaviour would be front-page news within an hour of its discovery. The plain truth is that even if your only motivation is to make profits, you're better off producing a safe and effective product that's trusted and approved the world over.

In the end, it's the very fact that vaccines are so useful and so profitable that offers the best protection against profiteering and corporate malfeasance. Just put yourself at the top of that gleaming glass building, looking out across the skyline as billions of dollars pour into your pharmaceutical empire. Why would you want to risk all

that for the sake of cutting a few corners here and there? The brutal yet reassuring truth about vaccines and big pharma is that there's just too much money to be made by doing things properly and by the book, so there's little incentive to put profits before people and risk bringing the whole thing crashing down.

Once viewed from this insider's perspective, it quickly becomes clear that it's the very same profit motive which some people decry that makes safe vaccines cheaper, more effective and more readily available with each passing year.

10

I'D RATHER BUILD NATURAL IMMUNITY

I have to confess that I was a little taken aback when this argument first emerged during a casual conversation, especially as that conversation was taking place over a dinner date with a girl whom I'm going to call Sandy. Yeah, I do take off the white coat every now and then, as there are some things in life that you just can't do on a blackboard.

As we were talking about my work, I was happy to be handed a golden opportunity to impress Sandy with my big, beautiful and scientifically trained brain. We all like to play to our strengths, although this was one conversation that didn't go in the direction I expected at all.

When Sandy told me that she preferred to develop a strong natural immunity rather than get vaccination jabs, I responded by wading straight in with the same seatbelt analogy we've already discussed in an earlier chapter. It was only after we'd been talking for a while that I began to realise that perhaps she was expressing an idea I hadn't actually come across before, which is something of a rarity these days.

Sandy was a very accomplished and professional lady, with a well-paid and important job in high finance. She was a real go-getter, a woman who wanted to do her very best and who also wanted the very best for herself. I suppose it goes without saying that she was really bright, educated and had a pretty decent layman's understanding of what vaccines are and how they actually work.

She pointed out to me that vaccines, as we've already discussed, are in fact a weakened or sometimes inactive version of a potent pathogen such as chickenpox or influenza virus. She was well aware that vaccines are a method used to 'train' or 'educate' the human immune system to recognise these harmful viruses and fight them off much more effectively than it could otherwise do. Based on that fairly accurate understanding, Sandy reasoned that she didn't really see how the weakened and inactive pathogens contained in vaccines could teach her immune system anything useful. As far as she was concerned it was all or nothing, and she had no time to waste on half measures like wishy-washy, weakened diseases with their hands tied behind their backs. If she caught something nasty, she was fit enough to fight it off, and afterwards her body would be completely familiar with that live and hostile pathogen in its most ferocious and feral form.

I'll never forget the way she described wanting an Ivy League education for her immune system, rather than a run-of-the-mill community college degree afforded by weakened vaccines.

It's probably fair to say that Sandy's fast-paced, high-pressure and demanding work environment meant that she enjoyed a challenge and was a highly competitive person, both by nature and training. On reflection, I think maybe it's that kind of cut-throat, kill-or-be-killed, survival-of-the-fittest outlook which coloured her thinking on a number of different issues, including medicine and vaccines. Coming from that sort of background, it had to be nothing but the best for Sandy.

Having spent a number of years in the field of vaccine development, I'm seldom surprised about the things I hear when people talk to me about vaccinations, but I must confess that Sandy's outlook was kind of a curveball.

By now I guess you won't be surprised to learn that the 'social responsibility' argument also didn't really cut much ice with Sandy either. It's not that she thought vaccines were a pointless waste of time, in fact quite the reverse. It's just that she believed they were for the other, slower, older and weaker members of the herd rather than for the likes of her. There should be no half-measures for hard-charging alpha females like Sandy, and nothing could truly replicate the naturally unfettered and wild versions of those pathogens waiting out there in the microscopic world.

So, there I was, stranded in the middle of this surreal conversation about vaccines with nowhere else to go. Suddenly an idea occurred to me along the lines of the 'sliding doors' parallel universe.

For anyone who's not familiar with the term 'sliding doors', it's derived from the romantic comedy bearing the same name. Basically, *Sliding Doors* is a movie about people living out alternate timelines, or alternative destinies if you like. It's an idea that's pretty common in science fiction and fantasy, but a lot rarer in light-hearted, romantic, feel-good films like romcoms.

So, with renewed enthusiasm I set out to explain a hypothetical case of how things might work out if my date had taken the flu shot (influenza vaccine), as compared to how events might've unfolded if she'd gone the other way and refused the needle.

So, if Sandy chooses the first sliding door, she'll be riding the tube to work every day without vaccination protection. In this scenario, all we have to do is just wait for the flu season and see how long it takes before she's exposed to the virulent and contagious influenza virus. Take it from a professional; it wouldn't be very long at all. So, what happens when somebody who is not vaccinated is exposed to the flu virus?

Well, they get the flu.

So now that Sandy's got the flu, she's going to spend about a week in bed feeling pretty bloody awful, to put it bluntly. Assuming there are no serious complications, she'll miss work and will probably have spread the virus a fair way before she started showing any significant symptoms. Now it's true that by the end of the whole episode, Sandy will have developed a greater immunity against that particular strain of the influenza virus, just as she'd said. She could call that a win in a roundabout way, although it's one heck of a risk to take when you consider that's a best-case scenario based on assuming that she didn't experience any complications, and didn't accidently pass the virus on to a vulnerable person and kill them.

However, if Sandy chooses the other sliding door and gets that flu shot, she'll still be riding the tube every day and will probably still be exposed to the virus during flu season. In this parallel timeline, however, her symptoms would be much more like that of a common cold. That's because her immune system has already been trained to recognise that particular flu virus strain and start fighting before it can take hold in her system. Sure, she might still miss a day of work, if at all, but she'll be a lot less infectious to those around her and much less likely to develop serious complications. The best part is that her immune system has still come across the real live virus, so she'll still end up reaching that peak immunity which is trained to recognise the full-blown, unadulterated and 'natural' virus strain. The big difference is that she has managed to reach the same destination while exposing herself, her career and her colleagues to a far lesser risk. Arguably, she could even end up with a stronger overall immune response, due to immune memory and the recall which has now occurred.

While Sandy was quite right to say that the virus in a vaccine is only a shadow of the real thing that waits out there in the real world, she was wrong to claim that it couldn't help her to develop a strong natural immunity. Vaccination could easily be described as a kind of training course for the body's immune system. When thought of in those terms, getting a flu shot isn't really that much different from learning to drive a car or perhaps training to perform a dangerous task like searching a smoke-filled building.

It's certainly true that practice tests and a real-life scenario are usually very different experiences, but would

anyone seriously suggest that driving tests should be abolished for that reason, or that firemen don't really require any training? What could possibly go wrong? I think the answer to that question is obvious to just about everyone.

While there's no substitute for real-world experience, there's also a lot to be said for training and preparation. If there wasn't, nobody would invest the time, energy and expense of doing it. This is just as true for the human body's natural immune response as for any of the other tasks we carry out and the roles we perform in our daily lives. Ultimately, if you're exposed to a pathogen in the real world, your body will try to build a strong immunity against it, but the outcome will be much improved if it's undergone some training before the event.

VACCINATIONS DON'T WORK/THEY'RE NOT SAFE

For the final chapter of our discussion, I thought it would be a good idea to wrap up by talking about a couple of the more generalised and poorly defined objections to mass vaccination that I often come across.

Non-specific statements like 'Vaccines don't work' or 'They're not safe' are probably the two most common starting points I've heard when I'm discussing someone's doubts. Although such a broad and sweeping objection might hide a more specific worry like the autism argument or fears about mercury, it very often boils down to just a general and ill-defined distrust of vaccinations that's not really based on any specific argument or piece of information, reliable or otherwise.

Since there are no particular, tangible ideas to get hold of and discuss in a situation like this, it's much more difficult to use illuminating analogies or fun facts to counter something that isn't really an argument at all. It's more like the shadow of an argument.

However, that's not to say that there's no reasonable and compelling counterargument to such vague and general feelings; it's just that the case for vaccination needs to be kept fairly broad and non-specific to mirror the objection.

Simplicity is the key in this kind of conversation, so I usually just point out a few general but indisputable facts surrounding the efficacy and safety of population vaccination.

For example, it's an inescapable fact that mass vaccination is the single most important and successful medical intervention ever devised.

It's also the safest . . . by far.

Of course, it's easy to just say that, but in the absence of any particular argument against vaccination, a few very general but indisputable and observable facts never go amiss.

Prior to mass vaccination, smallpox was one of the most deadly diseases known to mankind. In fact, it was so deadly that it was responsible for more than 300 million deaths in the twentieth century alone. It's worth pausing to just think about that staggering and almost unimaginable statistic for a moment. That's nearly the equivalent of the entire current US population, killed by a single infectious disease within living human memory.

Vaccination has not only reduced this number to zero, but it has also eradicated the scourge of smallpox from the entire planet. That is something which medical science has never achieved before in all of recorded human history. The killer of 300 million people has been permanently eliminated as a threat to human health, and all because of a simple vaccination injection.

Okay, so smallpox is the exception in having been completely wiped out, but other once common killers are now virtually unknown here in the developed world. The fact that polio is almost unheard of today is not some random but fortunate turn of events; it is the direct result of mass vaccination.

While it's not always helpful to start quoting statistical tables to support the undeniable benefits of vaccines, I always suggest that anyone who doubts their efficacy takes a good long look at the world around them as it is today, and then glances at a medical history book. The staggering effect of vaccination on human health should quickly become self-evident.

Turning to safety, there's a good chance that someone who claims that vaccines simply aren't safe is thinking about the fake autism scare, or maybe they've heard something about mercury in vaccines. If that's the case, we've already covered those subjects in some detail earlier on. However, just like in the case of reliability, there are some people who just have a general feeling that vaccines are somehow unsafe, or at least a good deal less safe than the medical establishment is letting on. This idea may not have developed from any specific information, but could have instead grown from a series of conversations they've heard and Internet articles they've seen over a number of years.

The truth of the matter is that vaccines are safe, and that serious adverse reactions are extremely rare. To put that in some sort of context, World Health Organisation figures show that no fewer than 116.5 million infants were successfully vaccinated worldwide in 2016 alone. Given the enormous scale of these mass vaccination programs, is it really a credible idea that governments and special interests are successfully covering up some systematic problem connected with modern vaccines?

Just ask yourself the following question: How likely is it that an unsafe product could be administered to 116.5 million kids in a single year and nobody would notice? Or, to ask a slightly different question, how likely is it that literally thousands of doctors and health professionals around the world would be either unaware of or actively supressing adverse vaccination data?

As discussed previously, that's even less likely than thousands of people deliberately keeping quiet about a supposedly fake moon landing. After all, a fake moon landing is no threat to human health.

Back down here in reality, mass vaccination programs have been operating in many diverse nations for over fifty years now, and there have been no serious or credible systemic health problems reported in all of that time. In fact, it's worth dwelling on that fifty-year figure for a little while, because it really helps to illustrate the point.

State-sponsored, mass vaccination has been a reality of life in many nations for the latter half of the twentieth century, and a good deal longer in some cases. During that time, a large number of diverse and often competing nations and cultures have come to embrace the social and individual benefits associated with these programs. The capitalist West thinks vaccination is a good idea and the Soviet Union thought it was a good idea too. What's more, those former Soviet bloc nations still continue the practice long after the Cold War has ended. Mass vaccination has been and continues to be practised from the frozen, northernmost tip of Alaska, right through western, central and eastern Europe. There's also modern Russia and its satellite states, down through China and into the Middle East, India and the Far East too. With an ever-growing number of competing countries and cultures promoting mass vaccination, there would be absolutely no reason for all of them to continue such a practice if it was thought to be generally unsafe.

What would communist China or the old Soviet Union stand to gain by first adopting and then continuing a medical practice which was mostly developed in Europe and the United States ... unless it was overwhelmingly beneficial medically, socially, economically and politically? The only reason a number of competing and even hostile nations would all continue doing basically the same thing is that they all agree that the case for doing so is overwhelming.

The fact of the matter is that vaccines are undeniably safe and effective, despite what someone may have heard somewhere once, or what other vague misgivings they might secretly be nursing. That simple statement is supported by masses of data stretching back for decades, and in some cases the better part of a century. If there was something fundamentally unsafe or ineffective about vaccines, we'd have noticed it long ago, and some doctor somewhere would have gotten extremely rich and famous by pointing it out. It is also important to notice that this book and its arguments are about modern-day vaccines; I am well aware of the horror stories and mistakes of history that happened in the early days of vaccination. However, arguing that vaccines are unsafe based on the state of the art from a half century ago is like arguing that travelling by plane is dangerous based on the large number of crashes and deaths in the 1940s.

Naturally, while this is all helpful information, we talked earlier about the human mind being generally bad at thinking statistically. Or in other words, we might be able to understand numbers and probabilities rationally, but we fail to connect those abstract mathematical ideas to our everyday experiences.

A constant theme throughout our discussion has been that it's always best to relate any data to a real-life situation, or to personalise the abstract if you want to put it another way.

In the case of someone who kind of just feels that vaccines aren't very safe, then a few simple questions can normally trigger some interesting discussion.

Have you ever met anyone who has had a serious adverse reaction to a vaccine? I don't mean heard about, and I don't mean read somewhere on the Internet either; I mean met, in your personal, real-life experience.

With vaccination rates at over 90 per cent in the developed world, the chances of meeting someone who has reacted badly to a vaccine would be very high indeed if they were somehow unsafe.

The same simple reasoning follows for effectiveness too, because if vaccines didn't work all that well, then there's a good chance that any given doubter would've met a victim of poliovirus or German measles. The fact that such an occurrence is quite unlikely today only goes to prove that vaccines must be effective, because these once common diseases are now virtually unheard of in the early twenty-first century.

While there are so many apparently unsolvable problems afflicting our world, it's comforting to think that diseases which once seemed invincible are now eradicated, or so unusual that we no longer see their devastating effects with our own eyes.

There's so much in this life that we can neither predict nor control, but thanks to vaccines, we have at least freed ourselves from the constant fear of many once common communicable diseases, and that's something you literally can't put a price on.

At the end of the day, if you don't trust the medical establishment or the pharmaceutical companies, then just look around you. Most people you know and have ever met

have been effectively protected from the diseases they were vaccinated against, and it's hugely unlikely that any of them have experienced some sinister side effect. Just relying on what you can see with your own eyes should provide you with some real-world perspective on the issue. In all likelihood, the majority of your schoolmates, workmates, friends and family members have been immunised at some stage without anything bad happening to them.

Just ask yourself this final question around the subject. What do you trust more: the evidence of your own eyes, or a bunch of anti-vaxxer Internet memes dreamt up by total strangers?

Made in the USA
Middletown, DE
07 August 2024

58670643R00054